AF811387

THE EXIGENCIES OF GLOBAL, INTEGRATED WARFARE: THE EVOLVING ROLE OF THE CJCS AND HIS DEDICATED STAFF

Michael S. Bell

May 2004

*****

The views expressed in this report are those of the author and do not necessarily reflect the official policy or position of the Department of the Army, the Department of Defense, or the U.S. Government. This report is cleared for public release; distribution is unlimited.

*****

Comments pertaining to this report are invited and should be forwarded to: Director, Strategic Studies Institute, U.S. Army War College, 122 Forbes Ave, Carlisle, PA 17013-5244. Copies of this report may be obtained from the Publications Office by calling (717) 245-4133, FAX (717) 245-3820, or by e-mail at *Rita.Rummel@carlisle.army.mil*

*****

All Strategic Studies Institute (SSI) monographs are available on the SSI Homepage for electronic dissemination. SSI's Homepage address is: *http://www.carlisle.army.mil/ssi/*

*****

The Strategic Studies Institute publishes a monthly e-mail newsletter to update the national security community on the research of our analysts, recent and forthcoming publications, and upcoming conferences sponsored by the Institute. Each newsletter also provides a strategic commentary by one of our research analysts. If you are interested in receiving this newsletter, please let us know by e-mail at *outreach@carlisle.army.mil* or by calling (717) 245-3133.

ISBN 1-58487-157-1

FOREWORD

The post-9/11 security environment and the demands of global, integrated operations and campaigns in a long-term war on terrorism underscore the need to consider new ideas that enhance the effectiveness of the military instrument of power while recognizing the inherent value of the existing system.

In order to better understand the character and enduring attributes of the position of Chairman of the Joint Chiefs of Staff and inform future Defense reform initiatives in the post 9/11 era, Lieutenant Colonel Michael S. Bell traces the Chairman's evolving role since the inception of the position during the Second World War through the Goldwater-Nichols reforms of the 1980s. Although Defense reformers often focus on more efficient business and budgeting practices, his narrative compels greater consideration of the value of apolitical military advice, civilian direction of policy, and legislative oversight on the military instrument of power. The position of the Chairman of the Joint Chiefs of Staff, supported by a dedicated Joint Staff, remains relevant and crucial in a security environment where technology is extending the capabilities and reach of both state and non-state actors. Arguably, the need to transcend a single service, capability, or regional perspective is even more essential today than it was when Congress formulated Goldwater-Nichols almost twenty years ago.

The Strategic Studies Institute is pleased to offer this monograph as an important contribution to an informed debate on the vital subject of Defense reform.

DOUGLAS C. LOVELACE, JR.
Director
Strategic Studies Institute

BIOGRAPHICAL SKETCH OF THE AUTHOR

MICHAEL S. BELL, a Lieutenant Colonel in the U.S. Army, is a strategist in the Directorate of Strategic Plans and Policy, the Joint Staff. Prior to this assignment, he commanded the 1st Battalion, 8th Cavalry, and has served in a variety of assignments in Europe, Southwest Asia, and the Continental United States. A former assistant professor of American history at the U.S. Military Academy, he holds an M.A. from the University of Maryland, recently defended his doctoral dissertation on the strategic views and perspectives of Franklin D. Roosevelt, and will be awarded a Ph.D. in the summer of 2004.

SUMMARY

Professional military advice from the Chairman of the Joint Chiefs of Staff (the Chairman or CJCS), informed and supported by an independent Joint Staff, is more important than ever in the conduct of global, integrated operations after 9/11. For more than 60 years, the Chairman has played a vital role by providing military advice to the President, the National Security Council (NSC), the Secretary of Defense, and the Congress within the context of civilian control of the U.S. armed forces. The advice of the Chairman consists of much more than his personal views and opinions; it represents the synthesis of the broad operational experience, military judgment, and technical expertise found in a Joint Staff dedicated to the Chairman.

Today, the United States has entered a new phase in the history of the Republic, and the armed forces have embarked upon the initial campaigns in a worldwide conflict against terrorism. The exigencies of the post-9/11 security environment and the demands of a long-term global war underscore the need to broaden and formalize the operational and supervisory responsibilities of the Chairman while retaining the fundamental, enduring character of the current system. In the prosecution of a global war that demands a range of professional military advice and insight from a strategic, joint, and integrated perspective, it is essential to retain an independent staff to assist the CJCS in formulating his national security input to the President, the NSC, and the Secretary of Defense, and in providing his strategic direction to the armed forces. The Chairman, furthermore, should be designated as the principal military advisor to the Homeland Security Council and entrusted with responsibility for supervising the Combatant Commanders and integrating and synchronizing their regional efforts with the actions of other government agencies into a global campaign.

THE EXIGENCIES OF GLOBAL, INTEGRATED WARFARE: THE EVOLVING ROLE OF THE CJCS AND HIS DEDICATED STAFF

THE EVOLUTION OF THE CJCS AND THE JOINT STAFF

The independent and advisory role of the Chairman of the Joint Chiefs of Staff (CJCS), codified in the Goldwater-Nichols Defense Reorganization Act and Department of Defense (DoD) Directive 5100.1, rests on over 60 years of practice and long-standing tradition in the United States.[1] Based on lessons and perceptions from the Revolutionary War through World War I, grounded in the early months of World War II, and reinforced during subsequent administrations, the CJCS has served as the senior military advisor to the Commander-in-Chief to coordinate and control the efforts of the armed forces. Beginning as an informal personal staff during World War II and formally established in 1947, the Joint Staff evolved and developed into a complex and capable organization to assist the CJCS in formulating advice and recommendations. At the same time, command of U.S. military forces remained in the hands of civilian leadership: the President and, after 1947, the Secretary of Defense.

In July 1939, expecting the outbreak of war in Europe, Congress gave President Franklin D. Roosevelt the authority to set up the executive agencies he believed essential to the defense of the United States. With the creation of the Executive Office of the President, Roosevelt placed the offices of the Chief of Naval Operations and the Chief of Staff of the Army, to include the strategic planners of the Army and the Navy, under his direct control.[2] Then in July 1942, he brought Admiral William D. Leahy out of retirement to preside over the wartime Joint Chiefs of Staff.[3] In terms of relative rank, Leahy was the most senior of the five-star general and flag officers appointed in 1944.[4]

Leahy's tenure established the heritage of independent advice from the Joint Chiefs of Staff (JCS) while retaining civilian control of the U.S. military establishment. During World War II, Leahy functioned with a staff of personal assistants and a group of officers to man the White House Map Room.[5] During the war, the staff supporting Leahy and the collective Joint Chiefs was further augmented by the creation of joint working groups and the joint Army-Navy strategic planning structure.[6] During World War II, the value of a dedicated staff to support Leahy was evident, and the National Security Act (NSA) of 1947 formally established the Joint Staff and authorized a complement of 100 officers.[7] After Roosevelt's death, Harry S. Truman continued the existing, informal relationship and retained Leahy as his advisor during his first term.

Following Leahy's retirement, Truman and Secretary of Defense James Forrestal asked General Dwight D. Eisenhower, then the president of Columbia University, to serve as the informal "presiding officer" of the Joint Chiefs. Forrestal wanted Eisenhower to coordinate the Services and present unified military advice, a role not covered in the 1947 NSA. By late 1948, Forrestal recognized the need for "a chairman with power to make specific recommendations to the Secretary of Defense." He thought that by bringing in Eisenhower "to advise and consult with me" that he would be able to get the position formalized. In January 1949, Eisenhower returned to active duty as Forrestal's "principal military advisor." Forrestal asked Eisenhower to help formulate amendments to the 1947 NSA, to produce a war plan that could form the basis for future military budgets, and to resolve strategy, service rivalries, and fiscal limitations.[8]

In early 1949, amid bitter interservice rivalry, Eisenhower advised President Truman that he needed "to appoint a president of the chiefs of staff and assign him to Forrestal." After a long conversation with the President on February 9, 1949, Eisenhower "agreed to act as chairman of joint chiefs of staff for a brief (I hope) period, pending change of law or formal arrangements for getting 'unification' on the rails." Clearly, the key for Eisenhower was to enact an enduring system that would continue to function irrespective of the particular strengths, weaknesses, and experience of either military advisors or civilian leaders.[9]

With Eisenhower's urging, Truman recognized the need to formalize the position of CJCS and to give the position enduring authority, no matter who filled the role. Truman agreed that the position had to be formalized by either law or appointment.[10] The 1949 amendment of the NSA codified the position and advisory function of the Chairman. It also more than doubled the size of the Joint Staff to 210 officers.[11]

After taking office in 1953, President Eisenhower chartered a Defense reorganization study headed by Nelson A. Rockefeller. Eisenhower accepted the recommendations of the Rockefeller Committee to include strengthening the staff of the Secretary of Defense, and at the same time making the CJCS responsible for managing the work of the Joint Staff and broadening the role of the Joint Staff in strategic planning. Eisenhower's initial reforms became effective in mid-1953.[12] Eisenhower's subsequent Defense reorganization proposals in 1958 further strengthened the authority of the Secretary of Defense and the role of the Chairman. Eisenhower advocated vesting the Secretary of Defense, rather than the Services, with the sole legal responsibility for combat operations, while empowering the CJCS, supported by an enlarged Joint Staff, to assist the Secretary of Defense in controlling the new unified commands. The resulting legislation nearly doubled the size of the Joint Staff to 400 officers.[13]

Consistent during both the Truman and Eisenhower administrations was the sense that they faced a new and changed security environment. The power of atomic weapons and aggressive Soviet Communism transformed Truman's world. Eisenhower characterized the threat posed in the 1950s by ballistic missiles and the revolutionary ability of adversaries to harness "an advancing industrial, military, and scientific establishment" as "unique in history."[14] Both presidents, despite their own substantial experience, recognized the value of independent advice and institutionalized the role of the CJCS supported by a Joint Staff to assist decisionmakers in dealing with a new world. Eisenhower asserted that his reforms would produce a more responsive and "accelerated decisionmaking process" that will better prepare "our country to meet an emergency which could come with little warning."[15]

The next period of Defense reform impacting the duties of the CJCS and the Joint Staff occurred in the 1980s. In 1982, the House

Committee on Armed Services proposed several changes designed "to improve the efficiency and effectiveness of the Joint Chiefs of Staff." In early 1982, Secretary of Defense Caspar Weinberger, CJCS General David C. Jones, and Chief of Staff of the Army General Edward C. Meyer criticized the existing organization of the JCS. The resulting legislation in the House of Representatives called for the creation of a Deputy Chairman, authorized the CJCS "to provide military advice in his own right," and stipulated that the Joint Staff be "independently operated" under the management of the Chairman.[16] The House legislation did not pass the Senate, and congressional hearings on Joint Chiefs of Staff reform and Defense reorganization continued for the next few years, culminating in the 1986 Goldwater-Nichols Act.

Congressional leaders and their outside experts who framed Goldwater-Nichols intended for the legislation to fix several perceived problems concerning the roles and responsibilities of the CJCS. They believed that the system hampered the ability of the Chairman to exercise independent authority or present much more than his personal views on joint issues affecting more than one Service. Rather than offering the President, the National Security Council (NSC), and the Secretary of Defense a valuable array of policy alternatives, reformers also assessed that elaborate Pentagon staffing procedures fueled a desire for unanimity among the JCS and effectively reduced their advice to the least common denominator. The result was military advice that was neither useful nor timely, degrading the power and influence of the Secretary of Defense. The framers concluded that the quality of advice the Chairman provided was personality dependent and hindered by the requirement for the Joint Staff to support all the JCS.[17]

In assessing the problem, Congress also disparaged the growth in the role of civilian advisors to the Secretary of Defense relative to the diminution of the role of military advisors and the Joint Staff. They observed that historically Secretaries of Defense often "turned to the mostly civilian staffs in the Office of the Secretary of Defense (OSD) for the kind of military advice they should have received from the JCS." Congress noted that a recent presidential commission had come to the conclusion that the President and Secretary of Defense "require military advice" that integrates the views of the Services and the combatant commanders and draws "upon the best thinking of . . . our senior military leadership." The contemporary weakness of the Joint Staff, constrained by statutory limits on its size and reliant on the staffs of the Services for analytical support, contributed to the tendency of the Secretary of Defense to turn to his civilian staff.[18]

During the Defense reorganization debate in 1986, President Ronald Reagan provided Congress with his assessment of the indispensable role of the CJCS and the need for any organizational reforms to transcend the particular strengths and weaknesses of serving individuals. Citing "rapid changes in the military challenges we face," Reagan stressed the need for a "highly adaptable" defense establishment that could "respond successfully" in a wide variety of changing circumstances and environments. Rather than structural changes that would minimize the role of the Chairman or the Joint Staff, he advocated a more clear delineation of responsibilities. The Chairman's advice remained crucial for Reagan to enable him to respond to what he saw as a rapidly changing security environment. Reagan believed that "clear and unambiguous" roles and responsibilities could "establish sound, fundamental relationships among and between civilian and military authorities" without limiting operational flexibility or "common sense" to deal with a wide variety of circumstances. He cautioned, "Laws must not be written in response to the strengths and weaknesses of individuals who now serve."[19]

The resulting Goldwater-Nichols Act designated the Chairman of the JCS as the principal military advisor to the President, the NSC, and the Secretary of Defense in order to strengthen civilian authority in the Department of Defense. The Act replaced the requirement for advice from the collective Joint Chiefs with advice from the Chairman of the JCS, allowing the other members of the Joint Chiefs to offer their guidance in agreement or amplification of the Chairman's position.[20]

The framers of Goldwater-Nichols intended that the Chairman be empowered to provide independent guidance even if that advice did not necessarily reflect unanimity with the other members of the JCS. They assessed that such advice was crucial to strengthening the effectiveness of the Secretary of Defense and the NSC by providing direct, uncompromised military guidance that did not rely on consensus and deal making that previously had characterized some deliberations of the JCS. To do so, the Chairman required the dedicated support of a staff that could conduct detailed, independent assessments and thorough analysis, reducing the influence of the Service staffs in the implementation of the Secretary's strategy and in the development of programs, budgets, and operational plans.[21]

The Congress believed that implementing Goldwater-Nichols would usher in a "revitalization of professional military advice." Congress recognized, however, that the quality of that guidance was contingent on the Joint Staff having two vital characteristics. The first was maintaining the independent role of the Joint Staff by placing it solely under the authority of the Chairman rather than the collective JCS. Second, assessing that the CJCS needed a "strong" and "substantial staff" comprised of "the broadest possible range of military experience and expertise," Goldwater-Nichols also revised the statutory cap on the size of the Joint Staff from 400 officers to 1,617 officers and civilians, and made joint experience a requirement for promotion to general or flag officer. By emphasizing that the Secretary of Defense "rely" on the JCS, assisted by an improved Joint Staff, "for staff support on military matters," the congressional framers of Goldwater-Nichols envisioned that future civilian decisions would be strengthened.[22]

The current system of professional military advice from the CJCS, supported by a dedicated staff, remains fundamentally sound and relevant. The system is enduring; since World War II, it has generated and continues to develop quality advice irrespective of personality or administration. The system provides integrated, joint, and global insights and advice to the President, NSC, and the Secretary of Defense. It deliberately preserves the principle of civilian control and direction while fostering an atmosphere of apolitical military advice and recommendations. Likewise, it underpins the process of checks and balances in the U.S. Federal Government by consistently providing the Congress with the professional military reports, information, and insight that it requires to implement effective legislative oversight of the Defense establishment. The Chairman's unique perspective transcends operational or regional boundaries and facilitates an unprecedented merging of joint

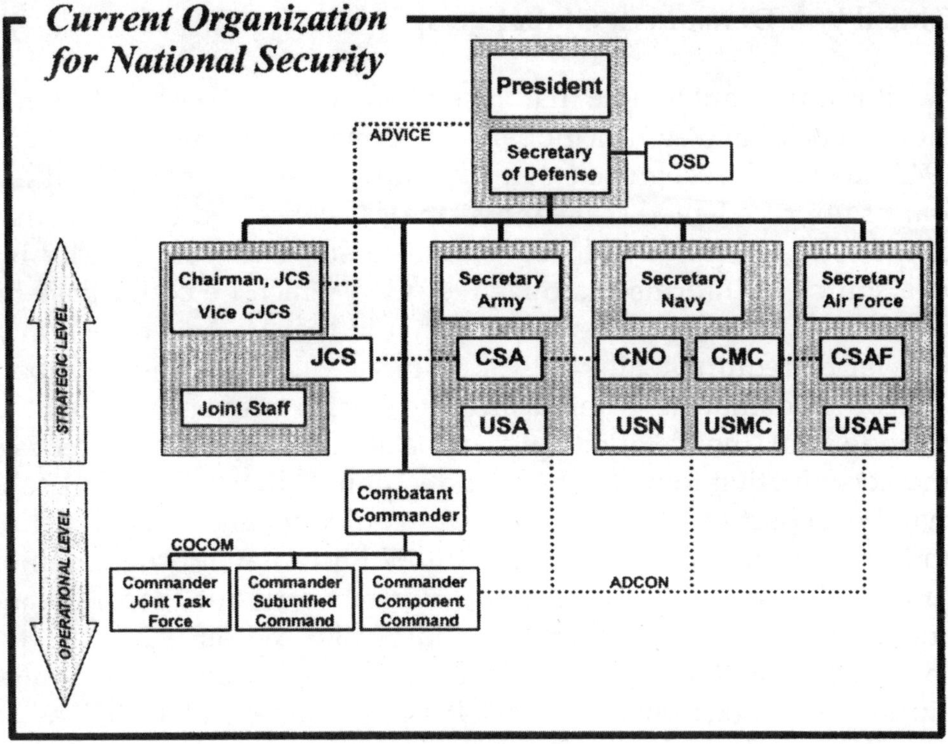

Figure 1.

capabilities and the integration of operations and activities across the U.S. Government.

In the post-9/11 world marked by the emergence of nonstate threats and a Global War on Terrorism, the Chairman provides a dedicated perspective well-suited to monitor and respond to transnational adversaries, empowered by communications and mobility systems and increasingly able to intercoordinate their activities with other adversaries on an unprecedented scale. The CJCS is unique in that he is the only senior military advisor in uniform who simultaneously possesses a dedicated strategic, global, and joint perspective. The Chairman transcends the regional or parochial perspectives of combatant commanders and the Service chiefs. The supporting personnel and systems of the Joint Staff allow the CJCS to maintain that perspective and provide quality professional advice that augments his personal insights and assessments.

Global War Demands a Global Perspective.

It is important to note that the position of CJCS was a solution to the exigencies of waging a global war against totalitarianism. In early 1942, Roosevelt recognized that the United States was involved in a war of unprecedented scale and scope. He conceived the Chairman of the Joint Chiefs would not serve as a supreme commander but as a source of informed knowledge upon which he could base his decisions. The President explained that in the strategic direction of the war he required "all kinds of opinions" and "finding out about things," and that he wanted an officer with proven judgment to provide him the detailed and painstaking "leg-work," analysis, and coordination to make the best decisions.[23] Roosevelt told Leahy that he expected the appointment would enhance his direction and coordination of the war and allow him to avoid the common strategic "mistake" of viewing the war as a series of "separate and unrelated" geographic fronts or consider only "the 'air war' . . . the 'land war' or the 'sea war'" in isolation.[24] Columnist Walter Lippmann characterized Roosevelt's appointment of Leahy as an organizational solution to the problems inherent in running a global war. Lippmann observed that, by "equipping the commander in chief with the resources that will enable him to make sound decisions in a global and three-dimensional war," Leahy enabled Roosevelt to make decisions on more than just "tidbits of information and intuition."[25] Secretary of War Henry L. Stimson concurred that the arrangement overcame Roosevelt's "weakness for snap decisions" while reinforcing his "sound strategic instincts."[26]

President Eisenhower had a similar need to support his national strategy for waging a potential war against the Soviet Union. Eisenhower sought to establish a series of interlocking regional alliances, patterned after the North Atlantic Treaty Organization (NATO), around the periphery of the Soviet Union. In addition to defending the countries of Europe, the Middle East, Southeast Asia, Japan, and Korea from Soviet aggression, the administration conceived that the alliances would provide bases for aerial reconnaissance operations and strikes against the Soviet Union.[27] Under Eisenhower, U.S. military actions around the world were considered part of a strategic design to enable the United States to prevail in a global war.

Eisenhower complemented his strategy with Defense reforms. Although he possessed exceptional strategic and operational military experience, Eisenhower's reforms increasing the size of the Joint Staff and vesting it with an integrated operations division were crafted to assist and better enable him "to see the totality of the national and international situation." Arguing that "separate ground, sea, and air warfare is gone forever" and that future war will involve "all services, as one single concentrated effort," Eisenhower wanted to institutionalize an independent staff that could provide the "professional military assistance required for efficient strategic planning and operational control."[28]

Discarding existing limitations on the size of the Joint Staff, Eisenhower envisioned an enlarged Joint Staff that would have the breadth and expertise to fill the expanded strategic role. An enlarged staff, he believed, would "broaden the degree of active participation" and "bring to bear more diversified and expert skills," thereby improving the quality of military advice and his range of strategic options.[29] His specific objective was to create a Joint Staff made up of officers who could transcend service parochialism and "center their entire effort on national planning for the over-all common defense of the nation and the West."[30] The result, from Eisenhower's perspective, enhanced his own power and ability as Commander in Chief. It is illustrative that Eisenhower singled out two interrelated aspects of the final 1958 Defense Reorganization Act to praise: "a reality of civilian control by the President and the Secretary of Defense" and the professional expertise provided by the Joint Chiefs of Staff and an enlarged Joint Staff.[31]

Although Eisenhower prepared for global war, the Cold War never devolved into open, direct conflict between the United States and the Soviet Union. Strategic deterrence and détente reduced the potential for global military confrontation between the superpowers. The challenges confronting Eisenhower's successors in Southeast Asia, the Middle East, Central America, and the Caribbean consisted of regional wars and insurgencies and small-scale contingency operations.

By the mid 1980s, the need for global strategic direction and coordination seemed to have diminished. Reflecting the contemporary mood, RAND Corporation analyst Francis Fukuyama rejected "the globalism of the 1950s." Rather than global war, Fukuyama asserted, the form of future conflict would require the United States to exercise the "prudent and selective application of military force" in discrete actions in the Third World.[32]

In 1986, the statements of President Reagan reflected the intellectual shift away from the need to prosecute and coordinate a global war to a CJCS role tailored to meet the need to respond effectively to regional contingencies and crises. During the debate over Defense reorganization, Reagan specifically urged strengthening "the ability of the military establishment to provide timely and integrated military advice to civilian leadership" by designating "the Chairman of the Joint Chiefs of Staff as the principal military advisor to the President, the Secretary of Defense, and the National Security Council." Rather than conceiving a need for strategic oversight to prosecute global war, however, Reagan concluded that the "special role" of the Chairman would enable him to provide a broad range of important advice "on operational military matters" and "joint military perspectives on both resource allocation and operations." In order to develop the best response by the United States when faced with a future crisis or contingency, Reagan assessed, "The highest quality military advice must be available to the President and the Secretary of Defense on a continuing basis" that provides "a clear, single, integrated military point of view" as well as "well-reasoned alternatives."[33]

Considering the trend of U.S. military operations since World War II, it is not surprising that the Goldwater-Nichols solution was to empower joint warfighters at the regional level. Because the collapse of the Soviet Union roughly coincided with the implementation of Goldwater-Nichols, operations in the decade and a half after 1986 have tended to validate its regional approach. Intellectually as well, the collapse of the Soviet Union seemed to herald the end of global conflict. Several years after the passage of Goldwater-Nichols, Fukuyama argued that the world was witnessing the fading of totalitarianisms and the emergence of "a remarkable consensus concerning the legitimacy of liberal democracy as a system of government." Fukuyama suggested the spread of liberalism and the victory of "the liberal *idea*" over other forms of government constituted the culmination or "end of history." Rather than any universal or global conflict, he surmised that any future challenges to liberal democracy would be local or regional in nature.[34]

The perceived elimination of any global challenge throughout the 1990s, amplified by the success of the regional Combatant Commands, fueled criticism that the large Pentagon bureaucracy was inefficient and too costly. As Defense reform advocates urged the Pentagon to adopt more efficient business practices, headquarters staffs became an obvious target. For example, James Locher, a Senate staffer during the Goldwater-Nichols debate, has argued that merging the military and civilian staffs in each of the military departments "could greatly improve efficiency and effectiveness." Merging most, if not all, of the Joint Staff into the Office of the Secretary of Defense would streamline processes and reduce the size of headquarters bloat. From Locher's perspective, the "duplicative structure, which originated in World War II, cannot be justified . . ."[35]

As a result, rather than acknowledge the strategic demands of coordinating and integrating global war, the call persists even after 9/11 for the Pentagon to adopt what have been characterized as cost effective business practices.[36] Rather than pursuing the goal of enhancing warfighting, reformers have targeted the size of headquarters bureaucracy, usually favoring the merger of major elements of the Joint Staff into OSD. Clearly, the product of such a merger would be a military establishment optimized to fight the peace operations and contingencies of the 1990s, not the global struggle of the future. Such a merged system, furthermore, has the potential to produce homogeneous strategic thinking and monochromatic military advice poorly suited to developing and analyzing the range of strategic options and alternatives required for waging campaigns against highly adaptive adversaries.

Although the Goldwater-Nichols regional solution proved effective prior to 9/11, it arguably fails to provide the optimal organizational structure to prosecute another global war, particularly one that has the potential to last for 20 or 30 years. Not since the end of World War II have the leadership of the U.S. Armed Forces had to synchronize the timing, sequencing, and priority of campaigns into a strategic design. Likewise, they have not been forced on a regular basis to allocate and balance finite resources between competing theaters simultaneously conducting vital combat operations.

The post-9/11 security environment warrants advice from a broad, global perspective complemented by an integrated, joint approach that synchronizes military activities and complements them with other instruments of national power, nongovernmental agencies, and allies. The inherent value of informed military advice remains inviolate. On a daily basis, the emerging strategic environment demonstrates the continued need for unvarnished and apolitical military advice, the same "frank" military advice that Roosevelt wanted to prosecute his global war.

Civilian Direction and Apolitical Advice.

At the onset of U.S. participation in World War II, Roosevelt acknowledged his need for professional military assistance to prosecute a global war and intentionally crafted a system that avoided the politicization of the American military while preserving informed civilian direction and decisionmaking authority. The military had the responsibility to execute established policy, to execute approved operations and campaigns, and to recommend options, changes, and alternatives. Roosevelt intended that the recall of Leahy to active duty would assist him in his personal direction and strategic conduct of the war. Decisionmaking authority remained exclusively in Roosevelt's hands; as Leahy noted, "*he* still was the Commander-in-Chief." Based on his understanding of the American political tradition and his observations from World War I, Roosevelt was adamant that the United States not create either a "Prussianized General Staff" or a French generalissimo that could conduct military operations or dictate policy independent of civilian control.[37] From Roosevelt's perspective, the concern was the ability of the "Prussian" or militaristic influences in the General Staff to step into the political arena and cower and overpower civilian leaders. He explained, "I have too much historical background and too much knowledge of existing dictatorships to make me desire any form of dictatorship for a democracy like the United States."[38]

Roosevelt's frame of reference was the situation in Imperial Germany and the European environment after World War I. Prior to the Great War, Kaiser Wilhem II acceded to his military leaders, and General Erich Falkenhaym, the supreme commander, also assumed the helm of the War Ministry.[39] For Roosevelt, the vice of militarism and the rise of fascism were the product of military involvement in political questions. Militarism, Roosevelt believed, was not confined to Germany alone but also had gained a hold in Italy and France.[40]

Closer to home, Roosevelt believed that militarism could be avoided in the United States by keeping civilian authority in a distinct channel, separate from an apolitical military.[41] Even before the outbreak of World War II, his administration distrusted the political aspirations of General Douglas MacArthur, the retired Army Chief of Staff who was considered "greedy for power." Reviewing the general's activities in the Philippines, Secretary of the Interior Harold L. Ickes assessed that MacArthur "comes pretty close to being a dictator."[42] Roosevelt considered MacArthur to be one of the two most dangerous men in the country. "We must tame these fellows," Roosevelt asserted, "and make them useful to us."[43]

Roosevelt's historical view of the forces affecting the early American Republic also shaded his perceptions of the danger of a politicized U.S. military. His views derived from his study of American history and his readings on Thomas Jefferson. For instance, in 1925, as he considered Jefferson's struggle between 1790 and 1800 against Federalists supported by the Army, he noted "the constantly recurring thought of parallel or at least analogous situations existing in our own generation." Roosevelt characterized Federalist Alexander Hamilton as a "convinced opponent of popular government," and "a virtual dictator" who with the support of the Army sought to destroy liberty in the United States. Roosevelt assessed, "I have a breathless feeling too as I . . . wonder if a century and a quarter later the same contending forces are not again mobilizing."[44]

Roosevelt's views captured a tradition of civilian control of policymaking deeply rooted in American civil-military culture.

In the colonial period, the actions of Oliver Cromwell, James II, and the garrisoning of British soldiers in North America after the Seven Years' War aroused fears of standing armies at the whim of an arbitrary and unchecked executive.[45] In 1776, George Mason drafted a Virginia declaration that proclaimed standing armies "dangerous to liberty" and cautioned "that in all cases the military should be under strict subordination to, and governed by, the civil power."[46] The Declaration of Independence echoed Mason's call for the military to operate under the oversight and authority of representative civilian assemblies. During the Revolutionary War General George Washington demonstrated his sense of responsibility to the Continental Congress and linked the patriotism and honor of Army officers with their deference to "the intentions of Congress."[47] Likewise, in his inaugural address, Thomas Jefferson praised "the supremacy of the civil over the military authority" as one of the basic principles that "guided our steps through an age of revolution and reformation."[48]

Presumably, additional strong influences on Roosevelt's thinking were the ideas of family friend Elihu Root, the Secretary of War during Theodore Roosevelt's administration. In 1903, Root had abolished the office of General Commanding the Army and replaced it with the position of Chief of Staff of the Army (CSA). Root also implemented a War Department General Staff to support the new CSA, a structure that remained through World War II. Root observed that it would not be suitable to have a General Staff organized exactly like either the German or French counterparts due to American political and military traditions.[49] Franklin D. Roosevelt's Secretary of War Henry L. Stimson praised Root's General Staff as adapting the "German invention" to fulfill "three requirements: civilian control in the executive branch, sound general planning, and constant cross-fertilization between the line of the Army and its high command in Washington."[50] The key aspect of Root's system of military control was to subordinate the Chief of Staff to civilian executive authority and have him act "under the directions of the President, or of the Secretary of War representing him." Of note is the fact that Roosevelt's designation of Leahy mirrored Root's language of "a military Chief of Staff to the President."[51]

When Harry S. Truman assumed the presidency, he retained Fleet Admiral Leahy as his Chief of Staff for nearly 4 years. Leahy's task was to keep the president "fully informed" and to provide frank advice. Although Truman demanded complete loyalty after a decision was made, he told Leahy that he expected immediate feedback if he thought the President was making a mistake.[52] Rather than an aide or crony, Leahy's role was to advise the President on military subjects, with some of those crossing into the political realm.[53] Truman was an admirer of the Roman general Cincinnatus who voluntarily had turned away from power, and his reforms embodied in the 1947 NSA reaffirmed civilian control of the military and placed atomic energy under a new civilian agency.[54] Recalling his decision to relieve General MacArthur of command in the Far East in 1951, MacArthur's correspondence with Truman's Republican critics incensed Truman, and Truman labeled the theater commander's public criticism of his national policy intolerable "insubordination."[55] Although Truman prized "how firmly the concept of the supremacy of the civil authority is accepted," he found it comforting to be advised by military "leaders of such ability and distinction."[56]

Eisenhower's reforms also strengthened civilian control while retaining the apolitical character of the CJCS and the Joint Staff. During the Defense reform debate in 1958, Eisenhower asserted that any reorganization required "a clear subordination of the military services to duly constituted civilian authority" that was "real" and not superficial.[57] He was adamant that the United States not create a "single chief of staff . . . Prussian staff" or "czar" able to conduct independent operations beyond the bounds of civilian oversight. Such an arrangement, Eisenhower asserted, had the potential to undermine civilian control and "threaten our liberty."[58]

Amid studies for Defense reorganization in 1986, President Reagan stressed the need for "the Chairman's exclusive control over the Joint Staff" in order to preserve the highest quality military advice and avoid politicizing the military. Rather than merge the Joint Staff with the OSD, Reagan concluded that the CJCS "should be supported by a military staff responsive to his own needs." Without a dedicated staff for analysis, the Chairman would be limited in the breadth and scope of advice he could provide civilian leaders.

The creation of a single, combined General Staff resulting from the merger of the Joint Staff into OSD, Reagan suggested, might have the consequence either of politicizing the military or impairing civilian authority for the proper functioning of the defense establishment.[59]

Reagan believed that the reforms he advocated would produce "civilian authority that is unimpaired and capable of strong executive action." He thought it appropriate to complement a more powerful Secretary of Defense with a stronger Chairman. He urged Congress to ensure that any Defense reorganization proposals preserve the special role of the Chairman and the voice of the collective JCS. In addition, he cautioned that any reorganization must guarantee that "the military establishment does not become embroiled in political matters." Reagan warned that aligning "the tenure of the Chairman or other senior officers" to changes resulting from the "civilian electoral process would endanger this heritage" and politicize the military establishment.[60]

Current proposals to merge significant portions of the Joint Staff into OSD ignore the cautions of Reagan. Reformers cite the British Ministry of Defence (MOD) as an example of a successful merged staff in which military officers and civilians serve side-by-side. The current MOD organization dates from 1984 and reflects the contemporary peacetime desire for greater financial and administrative economy.[61] The British MOD is a product of the same security environment that produced Goldwater-Nichols in the United States. The civilians in the Central Staff of the MOD, however, are civil servants, not political appointees. The integration of the officers and civil servants of the Joint Staff with OSD political appointees has the potential to politicize military advice or dilute frank assessments and feedback.

Checks and Balances on the Process of National Security.

Independent military advice from the CJCS, supported by the Joint Staff, is intended not only to assist the Secretary of Defense but also the President, the NSC, and the Congress. The continued availability of professional military advice and access to an array of integrated and joint options provides the President, the NSC, and the Secretary of Defense with a separate, dedicated and informed reservoir upon which to draw in support of their decisionmaking, particularly when time is essential. It also provides the legislative branch the detailed information required for effective congressional oversight.

Since the early days of the Republic, congressional oversight was seen as a prerequisite to keep the executive branch of the government and any standing military under the control of the people. During the debates of the Constitutional Convention, even the most ardent Federalists argued that congressional control over the army was necessary to ensure that the executive branch could not use the military improperly and overthrow the Republic or establish a dictatorship. Hamilton asserted that the United States needed a standing military, but that Congress "was not *at liberty* to invest in the executive permanent funds for support of the army, if they were incautious enough to be willing to repose in it so improper a confidence."[62] The balance that Hamilton proposed for the Constitution was giving the executive authority for "the supreme command and direction of the military and naval forces," while the obligation of the legislative branch extended "to the *declaring* of war and to the *raising* and *regulating* of fleets and armies."[63] Consistent with Hamilton's proposals, the Constitution drafted in 1788 reflected a system of checks and balances intended to prevent any one group or individual from securing too much power.[64]

Vested by the Constitution with the authority to raise and maintain an Army, Congress values the independent military advice of the Chairman. Congressional leaders respect the input and perspective of the CJCS, a trend that can be expected to continue as fewer and fewer elected officials have prior military experience. During recent confirmation hearings, the Senate Armed Services Committee (SASC) declared that they expected that the CJCS will continue to provide "testimony, briefings, and other communications of information" in order for Congress "to exercise its legislative and oversight responsibilities." Wanting more from the Chairman than purely formal input, congressional leaders have encouraged his frank and independent assessment when requested. For instance, during the reconfirmation of General Richard B. Myers, the SASC specifically asked the Chairman whether he would continue to provide his "personal views, even if those views differ from the administration in power."[65]

Ensuring the credibility of the Chairman's advice and personal views requires a process by which the CJCS can analyze the full range of national security issues and develop recommendations independent of the services and the staff of the OSD. As Truman recognized, such credibility is essential when the analysis and recommendations are both in agreement and in opposition to those of the Secretary of Defense and the President. The independent nature of the system through which the CJCS analyzes requirements and promulgates advice and recommendations increases the credibility of that advice. Military guidance requires operational experience and technical knowledge that political appointees or civil servants may not possess.

Goldwater-Nichols acknowledged that the best possible military advice from the Chairman should take a wide variety of forms to include "strategic plans, budget proposals, or joint training policies." As principal military advisor, Goldwater-Nichols envisioned CJCS participation in all major DOD decisionmaking processes, to include policy formulation as well as policy implementation and execution. It assigned the CJCS complementary and mutually supporting tasks in the areas of strategy, planning, programming and budgeting, and force employment. For example, statute requires the Chairman to conduct an independent assessment of the *Quadrennial Defense Review* conducted by the Secretary of Defense and an associated risk assessment of the Defense Strategy contained therein. He is also responsible for periodic readiness and risk assessments associated with the missions described in the current military strategy. The CJCS has provided advice in the form of alternatives to service and combatant command budget proposals and program recommendations that better conform with the priorities of the Secretary of Defense. Those independent assessments are designed to inform the Congress as well as the President and Secretary of Defense.[66]

Over the decade and a half since the passage of Goldwater-Nichols, quality advice from the Chairman rested on three fundamental and complementary attributes. One was a Joint Staff whose personnel possessed superior professional expertise and broad operational experience coupled with a deep and practical understanding of technologies, warfighting concepts, and force capabilities. The second aspect, and one which has complemented the expertise and breadth of the Joint Staff, has been a series of CJCS systems, programs, and processes that assess the global, strategic environment and analyze assigned tasks based on guidance from the President and the Secretary of Defense. Collectively, those collaborative processes have assisted the Chairman in his role as military advisor by transforming a spectrum of extremely diverse staff expertise, perspectives, and insights into specific advice and guidance for training and experimentation, assessments of current readiness levels, tracking current operations, capturing lessons learned, and crisis management guidance. The third aspect has been the level of objectivity provided by an independent staff under the authority, direction, and control of the Chairman. That arrangement has prevented the formulation of advice dominated by a single Service or driven by parochial or politically partisan views.

The overall product of those three attributes has been to enhance the quality of military advice through a dedicated joint and global perspective and one that increasingly integrates other government agencies. Because he remains focused on strategic military objectives, the Chairman's perspective is not constrained by advocacy of a specific capability or operational solution. What has resulted is multidimensional advice on issues affecting joint doctrine, operations, force structure and organization, professional education, joint training, personnel policies, budgets and programs, facilities and infrastructure, strategy, and risk determination and mitigation.

Goldwater-Nichols made the CJCS responsible for providing advice on the strategic direction of the Armed Forces. Congress intended that Goldwater-Nichols assign the Chairman with the specific duty "to help set military priorities in a fiscally constrained planning document" that was fully integrated and forward-looking.[67] It has fallen on the CJCS to examine policy and strategy recommendations, determine military implications, integrate those with other government agencies and with allies, and propose alternatives that meet national objectives but at reduced levels of risk. Collectively, the product-oriented CJCS processes generate the Chairman's policy formulation advice to the President, the NSC, and the Secretary of Defense on strategic direction and proposals for programs and budgets as well as policy implementation guidance to the Services and Combatant Commands on planning and force employment, to include training, doctrine, and crisis management.[68]

CONTINUED RELEVANCE AND THE EXIGENCIES OF THE FUTURE

By all accounts 9/11 marked the beginning of a new era, and the United States has embarked upon another global war. As such, it is useful to contrast the current era with the previous periods of Defense reform. What emerges is the awareness that, in each era, decisionmakers also believed that they faced unprecedented challenges to the security of the United States. Franklin D. Roosevelt waged a global war against the ideology of Nazism, fascism, and militarism equipped with technology that fundamentally affected the security of the United States. Roosevelt conceived that aircraft could strike the United States within hours.[69] For Truman and Eisenhower, the threat constituted Communist ideology embodied in an aggressive Soviet Union armed with nuclear weapons and ballistic missiles. Reagan envisioned a moral contest with an evil Soviet empire capable of harnessing rapidly changing military technology.[70]

While the demands of the War on Terrorism (WOT) will necessitate that the U.S. Armed Forces be better stewards of resources, arguably the fundamental goal of future Defense reforms should be to further enhance the strategic integration of military actions worldwide with the other elements of national power. Conceivably, the decisive, global application of military power in the WOT will require unity of effort and the central, strategic coordination of the actions of all military resources and regional forces to achieve the overarching goal of national survival. The problem is that the military command structure enacted by Goldwater-Nichols produced unity of effort at theater level while ignoring structures for strategic coordination and integration.

Between the regional Combatant Commands and the Secretary of Defense and the President, there currently is no military headquarters with authority to plan, coordinate, sequence, prioritize, and execute all aspects of the integrated, global fight. Although U.S. Strategic Command and U.S. Special Operations Command are globally postured Combatant Commands, each of those are limited in terms of perspective, expertise, and the assets they control; arguably, both would be predisposed to execute the global war using their discrete capabilities or assets. Another shortcoming is that the Goldwater-Nichols arrangement could place the conduct of military operations by combatant commanders outside the purview of the CJCS or any senior servicemember with the legal responsibility to provide military advice, potentially muffling the voices of senior officers and limiting the insights available to the Secretary of Defense, the NSC, the President, or the Congress. In contrast, the Chairman's unique perspective, not tied to a particular region or capability, becomes increasingly valuable against an elusive, global adversary able to adapt quickly and exploit existing regional and operational seams.

The Chairman remains uniquely positioned to assess strategic direction, readiness, and future risk across the joint force and the entire spectrum of military operations. Certainly, it is entirely appropriate for the CJCS to assist in providing for strategic direction of the armed forces and to assess impacts on the long-term health and readiness of the forces, and the levels of risk to the force associated with specific courses of action. Already, the Chairman's internal processes are under transformation to make them more adaptive and responsive. Nevertheless, two immediate legislative modifications are key to make the advice of the Chairman more relevant and anticipatory. First, Congress should amend the responsibilities of the Chairman in Chapter 5 of Title 10, U.S. Code, to designate him as the principal military advisor to the Homeland Security Council. Second, anticipation and responsiveness require information. Section 163 of Title 10, U.S. Code should be amended to direct that "communications between the President or the Secretary of Defense and the commanders of the unified and specified combatant commands *will* be transmitted through the CJCS." That amendment would merely formalize a practice followed in previous administrations. President Reagan, for instance, signed a directive implementing the practice of the Chairman functioning "within the chain of command by transmitting to the combatant commanders those orders I give to the Secretary."[71]

In addition to those two amendments, however, a progressive modification of the roles and functions of the Chairman and the Joint Staff demands serious legislative consideration. A compelling need exists for a single military officer supported by a dedicated military staff to be responsible for providing strategic direction and interagency integration of a global, joint, and combined WOT under the direction of the Secretary of Defense and the President.[72] Although such an evolution of the role of the CJCS could be carried out under the existing Title 10 authorities of the President and the Secretary of Defense who can vest the CJCS with operational responsibilities, legislative sanction would provide an enduring character beyond the term of an administration. The Chairman would continue to serve in his capacity as the principal military advisor, assisted by the other Joint Chiefs, but would also serve as the Chief of Joint Operations (CJO), the representative and executive agent of the Secretary of Defense and the President to supervise and direct the combatant commanders, to provide strategic direction for the joint force, and to coordinate operations preapproved by the Secretary and the President. The result would be civilian control, as Elihu Root characterized it, "exercised through a single military expert of high rank . . . who is bound to use all of his skill and knowledge in giving effect to the purposes and general directions of his civilian superior. . . ."[73] Commensurate with his increased role and supervisory authority over the four-star combatant commanders, the Chairman could be authorized the rank of Fleet Admiral, General of the Air Force, or General of the Army or Marine Corps.

To preserve the broadest possible perspective of military advice and Service expertise, the Services would remain as currently constituted, with the Service Departments subordinate to the Secretary of Defense rather than to the Chairman, and with the

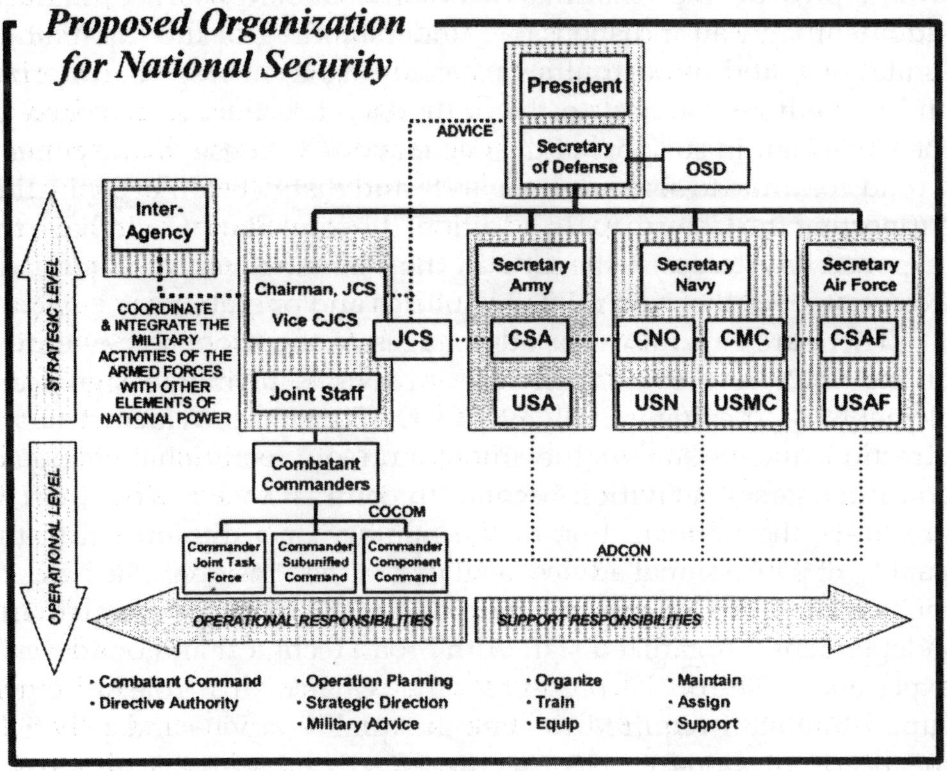

Figure 2.

Service chiefs remaining an integral part of the JCS. This would ensure the divergent viewpoints of the highest-ranking senior officers are not homogenized and remain available to provide civilian leaders with the information they require to organize, train, and equip forces for joint operations and to make the best national security choices for the United States.

The Chairman, dual-hatted as CJCS and CJO, would be supported by a Joint Forces General Staff vested with responsibilities in between the current scope of the Joint Staff and an overall Armed Forces General Staff.[74] This reoriented Joint Staff could be integrated along the lines of the Central Staff in the British MOD and similarly staffed with professional career civil servants and rotating military officers with technical skills and operational experience as is done today.[75] The rotational policy for uniformed staff officers would provide the Chairman with invaluable expertise, military judgment, and an indispensable understanding of the capabilities, limitations, and opportunities of emerging technologies, concepts, and procedures. Subject to the authority, direction, and control of the Chairman in support of the Secretary of Defense, the Joint Staff would continue to serve as a separate and distinct staff organization supporting the Chairman. In addition, the Joint Staff would have the responsibility to assist the CJO in the planning, coordination, and execution of authorized military policies and operations.

There are several major advantages of the proposed evolution of the CJCS role over alternative proposals to merge significant elements of the Joint Staff into OSD. First, it provides strategic direction and enhances the integration of operational, regional, and interagency activities. Second, in contrast to a merged staff, it precludes the politicization of the military or a diminution of the quality of professional advice available to the President, the NSC, or the Secretary of Defense by retaining for the CJCS a dedicated and independently organized staff with broad technical and operational experience. Third, it preserves the Chairman's role and the supporting staff required to generate quality advice and to satisfy the needs of Congress to execute effective legislative oversight. Neither wartime exigency nor administrative efficiency provides compelling justification to jettison the enduring system of checks and balances that underpin the American federal system. Finally, the proposal preserves balance in the DoD. The Secretary of Defense, by exercising command authority through the Chairman, would be able to dedicate greater energy to transformation, programming, and budgeting agendas. Although continuing to provide advice on programs and budgets, the primary focus of the Chairman would be strategic direction of operational warfighting.

The position of the CJCS remains relevant and crucial in today's security environment. Arguably, the need to transcend a single service or regional perspective is even more essential today than it was in the days of Franklin D. Roosevelt. Vital to the continued relevance of the Chairman to the future of the WOT is the modification of the role of the CJCS and the Joint Staff. Furthermore, in order to build enduring victories, the Chairman's unique global and joint perspective should be reinforced with the integration of interagency insights into operations from deliberate planning through execution. The Chairman can assist in the strategic integration of all elements of national power and provide the crucial linkage between the military activities of the WOT and other foreign policy and national security objectives at home and abroad.

Following 9/11, the United States has moved into a new era. The role of the CJCS should continue to evolve to meet the strategic demands of this new era while preserving the attributes of apolitical advice, civilian direction of policy, and legislative checks and balances on military and executive power. Defense reform and rationalization must ensure that the Chairman, supported by the Joint Staff, remains responsive and adaptive in order to serve the President, the NSC, and the Secretary of Defense in their prosecution of global war in the 21st century.

ENDNOTES

1. U.S. Public Law 99-433, *Goldwater-Nichols Department of Defense Reorganization Act of 1986*, Washington, DC: U.S. Government Printing Office, 1986 (commonly referred to as Goldwater-Nichols), Section 4.1, Department of Defense Directive 5100.1, August 1, 2002.

2. Eric Larrabee, *Commander in Chief: Franklin Delano Roosevelt, His Lieutenants, and Their War*, New York: Harper and Row, 1987, p. 16. Roosevelt transferred the Joint Army-Navy Board and the Joint Army-Navy Munitions Board to the newly created office.

3. Entry for July 18, 1942, William D. Leahy Diaries, and Franklin D. Roosevelt letter to Leahy, July 20, 1942, The Papers of William D. Leahy, Manuscript Division, Library of Congress.

4. Henry L. Stimson and James Forrestal memorandum for the President, December 13, 1944, OF-25, President's Official File, Franklin D. Roosevelt Presidential Library (hereafter FDRL), Hyde Park, New York.

5. Larrabee, pp. 16-21. Leahy had been recalled "to active duty as Chief of Staff to the Commander-in-Chief of the Army and Navy." Franklin D. Roosevelt memorandum to William D. Leahy, July 20, 1942, The Papers of William D. Leahy, Manuscript Division, Library of Congress. See also William D. Leahy, *I Was There: The Personal Story of the Chief of Staff to Presidents Roosevelt and Truman based on His Notes and Diaries Made at the Time*, New York: McGraw-Hill, 1950, p. 97.

6. Mark A. Stoler, *George C. Marshall: Soldier-Statesman of the American Century*, Boston: Twayne Publishers, 1989, pp. 102-103; Charles E. Kirkpatrick, *An Unknown Future and a Doubtful Present: Writing the Victory Plan of 1941*, Washington, DC: U.S. Government Printing Office, 1990, p. 51.

7. The National Security Act of 1947, July 26, 1947, in Alice C. Cole, *et al.*, *The Department of Defense: Documents on Defense Establishment and Organization, 1944-1978*, Washington, DC: Office of the Secretary of Defense Historical Office, 1978, p. 46.

8. James Forrestal, *The Forrestal Diaries*, Walter Millis, ed., New York: The Viking Press, 1951, pp. 539-540, 547-548.

9. Entry for February 9, 1949, Dwight D. Eisenhower, *The Eisenhower Diaries*, edited by Robert H. Ferrell, New York: W. W. Norton and Company, 1981, pp. 150-157.

10. *Ibid.*

11. Public Law 216, 81st Congress, August 10, 1949, in Alice C. Cole, *et al.*, *The Department of Defense: Documents on Defense Establishment and Organization, 1944-1978*, Washington, DC: Office of the Secretary of Defense Historical Office, 1978, pp. 94-95.

12. Dwight D. Eisenhower Message to Congress Transmitting Reorganization Plan No. 6, April 30, 1953, in Alice C. Cole, *et al.*, *The Department of Defense: Documents on Defense Establishment and Organization, 1944-1978*, Washington, DC: Office of the Secretary of Defense Historical Office, 1978, pp. 149-158.

13. Dwight D. Eisenhower, "The White House Years," Vol. 2, *Waging Peace, 1956-1961*, Garden City, New York: Doubleday and Company, 1965, pp. 221, 244-247, 251-252. Department of Defense Reorganization Act of 1958, August 6, 1958, in Alice C. Cole, *et al.*, *The Department of Defense: Documents on Defense Establishment and Organization, 1944-1978*, Washington, DC: Office of the Secretary of Defense Historical Office, 1978, pp. 188-230.

14. Dwight D. Eisenhower, "Annual Message to the Congress on the State of the Union," January 1958, *www.c-span.org/executive/transcript.asp?cat=current_event&code=bush-admin&year=1958* (link valid as of July 23, 2003). For details of Eisenhower's subsequent proposals, see Dwight D. Eisenhower message to Congress, April 3, 1958, in Alice C. Cole, *et al.*, *The Department of Defense: Documents on Defense Establishment and Organization, 1944-1978*, pp. 175-186.

15. Dwight D. Eisenhower message to Congress, April 3, 1958, in Alice C. Cole, *et al.*, *The Department of Defense: Documents on Defense Establishment and Organization, 1944-1978.*

16. United States, 97th Congress, House, Report No. 97-744, *Joint Chiefs of Staff Reorganization Act of 1982, Report to Accompany H.R. 6954*, Washington, DC: U.S. Government Printing Office, 1982, pp. 1-4.

17. United States, 99th Congress, Senate, Committee on Armed Services, *Defense Organization: The Need for Change: Staff Report to the Committee on Armed Services, United States Senate*, Washington, DC: U.S. Government Printing Office, 1985, pp. 171-178, 639-641.

18. United States, 99th Congress, Senate, Committee on Armed Services, *Department of Defense Reorganization Act of 1986 Report (To Accompany S. 2295) . . . Together With Additional Views*, Washington, DC: U.S. Government Printing Office, 1986, pp. 21-22, 31-32.

19. Ronald Reagan, "Message to Congress Outlining Proposals for Improving the Organization of the Defense Establishment," the White House, April 24, 1986, printed in United States, 99th Congress, 2nd Session, House Document 99-209, *Defense Reorganization: Message From the President of the United States*, Washington, DC: U.S. Government Printing Office, 1986.

20. United States, 99th Congress, Senate, Committee on Armed Services, *Department of Defense Reorganization Act of 1986 Report (To Accompany S. 2295). . . Together With Additional Views*, pp. 2-4, 21-22.

21. United States, 99th Congress, Senate, Committee on Armed Services, *Defense Organization: The Need for Change: Staff Report to the Committee on Armed Services, United States Senate*, Washington, DC: U.S. Government Printing Office, 1985, pp. 173-174, 209-224.

22. United States, 99th Congress, Senate, Committee on Armed Services, *Department of Defense Reorganization Act of 1986 Report (To Accompany S. 2295). . . Together With Additional Views*, pp. 22, 32-34.

23. Franklin D. Roosevelt, 836th Presidential Press Conference, July 21, 1942, Samuel I. Rosenman, ed., *The Public Papers and Addresses of Franklin D. Roosevelt, 1942: Humanity on the Defensive*, New York: Harper and Brothers, 1950, p. 302; William D. Leahy, *I Was There*, p. 97.

24. Entry for July 7, 1942, William D. Leahy Diaries; Franklin D. Roosevelt, Message to Congress on the Progress of the War, September 17, 1943, Samuel I. Rosenman, ed., *The Public Papers and Addresses of Franklin D. Roosevelt, 1943: The Tide Turns*, New York: Harper and Brothers, 1950, p. 394.

25. Walter Lippmann quoted in Paul L. Miles, "Roosevelt and Leahy: The Orchestration of Global Strategy," in Edward J. Marolda, ed., *FDR and the U.S. Navy*, New York: St. Martin's Press, 1998, pp. 151-152.

26. Stimson served as Secretary of War 1911-13, as Secretary of State from 1929-33, and as Secretary of War from 1940-45. Henry L. Stimson and McGeorge Bundy, *On Active Service in Peace and War*, New York: Harper and Brothers, 1948, p. 414.

27. Dwight D. Eisenhower, "The White House Years," Vol. 1, *Mandate For Change, 1953-1956*, Garden City, New York: Doubleday and Company, 1963, pp. 449-453, 574; John Lewis Gaddis, *Russia, the Soviet Union and the United States*, New York: McGraw-Hill, 1990, 2nd ed., pp. 229-230.

28. Dwight D. Eisenhower Message to Congress, April 3, 1958, pp. 175-186.

29. Dwight D. Eisenhower Message to Congress Transmitting Reorganization Plan No. 6, April 30, 1953, in Alice C. Cole, *et al.*, *The Department of Defense: Documents on Defense Establishment and Organization, 1944-1978*, Washington, DC: Office of the Secretary of Defense Historical Office, 1978, p. 155.

30. Dwight D. Eisenhower, "The White House Years," Vol. 1, *Mandate For Change, 1953-1956*, pp. 448-9.

31. Dwight D. Eisenhower, "The White House Years," Vol. 2, *Waging Peace, 1956-1961*, pp. 221, 244-247, 251-252.

32. Francis Fukuyama, *Military Aspects of the U.S.-Soviet Competition in the Third World*, Santa Monica, CA: RAND Corporation, 1985, p. 29.

33. Ronald Reagan, "Message to Congress Outlining Proposals for Improving the Organization of the Defense Establishment," the White House, April 24, 1986, printed in United States, 99th Congress, 2nd Session, House Document 99-209, *Defense Reorganization: Message From the President of the United States*, Washington, DC: U.S. Government Printing Office, 1986.

34. Francis Fukuyama, *The End of History and the Last Man*, New York: The Free Press, 1992; Avon Books, 1993, pp. xi, xxiii, 44-51, and 328-329.

35. James R. Locher III, *Victory on the Potomac: The Goldwater-Nichols Act Unifies the Pentagon*, College Station, TX: Texas A&M University Press, 2002, pp. 448-449. Locher began this book in 1993. See also Locher, "Has It Worked?: The Goldwater-Nichols Reorganization Act," *Naval War College Review*, Autumn 2001, Vol. LIV, No. 4, *http://www.nwc.navy.mil/press/Review/2001/Autumn/art5-au1.htm,* site valid as of October 21, 2003.

36. For example, the Washington-based Center for Strategic and International Studies (CSIS) is hoping to address perceived problems in "the efficiency and effectiveness of defense management" with the project that started in 2002. CSIS,

"Beyond Goldwater-Nichols: Defense Reform for a New Strategic Era," *http: //www.csis.org/isp/gn/*, link valid as of October 13, 2003.

37. William D. Leahy, *I Was There*, p. 97.

38. James A. Farley, *Jim Farley's Story: The Roosevelt Years*, New York: Whittlesey House, 1948, p. 128.

39. Gordon A. Craig, *The Politics of the Prussian Army, 1640-1945*, London: Oxford University Press, 1964, pp. 253-257, 291, 296-297, 303.

40. Hence his innate distrust of French generals and his particular antipathy for Free French leader Charles de Gaulle, who he assessed was "out to achieve one-man government in France." Elliott Roosevelt, *As He Saw It*, New York: Duell, Sloan and Pearce, 1946, p. 73. Roosevelt confided to his wife, "General de Gaulle is a soldier, patriotic, yes, devoted to his country; but, on the other hand, he is a politician and a fanatic and there are, I think, in him almost the makings of a dictator." [Anna] Eleanor Roosevelt, *The Autobiography of Eleanor Roosevelt*, New York: Da Capo Press, 1992, p. 248.

41. For example, Roosevelt assessed that Colonel Charles Lindbergh, a vocal isolationist and political opponent, had essentially become a Nazi. Following a 1940 radio address in which Lindbergh criticized the president's military preparedness program, Roosevelt complained that the aviator "has completely abandoned his belief in our form of government and has accepted Nazi methods." Entry for May 20, 1940, Henry Morgenthau, Jr., Presidential Diaries, FDRL.

42. Entry for January 30, 1937, Harold L. Ickes, "The Secret Diary of Harold L. Ickes," Vol. 2, *The Inside Struggle, 1936-1939*, New York: Simon and Schuster, 1954, p. 62. See also entry for August 6, 1940, "Joseph Lash Journal, 1939-42," Folder 3, Box 31, Speeches and Writings, Lash Papers, FDRL.

43. Rexford G. Tugwell, *The Democratic Roosevelt*, Garden City, New York: Doubleday and Company, 1957, pp. 349-350.

44. Typed copy of long-hand, "Review of Claude Bower's [sic] book, *Jefferson and Hamilton*, November 19, 1925," folder 64, Box 41, Writing and Statement File, Franklin D. Roosevelt: Family, Business, and Personal Papers, FDRL.

45. Bernard Bailyn, *Faces of Revolution: Personalities and Themes in the American Struggle for Independence*, New York: Alfred A. Knopf, 1990, p. 237.

46. The Virginia Declaration of Rights, 1776, reprinted in *Sources in American Constitutional History*, Michael Les Benedict, ed., Lexington, MA: D. C. Heath and Company, 1996, p. 20.

47. George Washington address to the officers of the Army, March 15, 1783, reprinted in *The Spirit of Seventy-Six,* edited by Henry Steele Commager and Richard B. Morris, New York: Bonanza Books, 1983, pp. 1283-1285.

48. First Inaugural Address, March 4, 1801, in Thomas Jefferson, *Writings,* Merrill D. Peterson, ed., New York: The Library of America, 1984, p. 495.

49. "Organization of the General Staff," in Elihu Root, *The Military and Colonial Policy of the United States: Addresses and Reports,* Cambridge, MA: Harvard University Press, 1916, p. 424.

50. Stimson had a friendship with Root that dated to 1893 when he entered Root's law firm in New York. Henry L. Stimson and McGeorge Bundy, *On Active Service in Peace and War,* New York: Harper and Brothers, 1948, pp. 32, xviii.

51. "Organization of the General Staff," in Elihu Root, *The Military and Colonial Policy of the United States: Addresses and Reports,* Cambridge, MA: Harvard University Press, 1916, p. 427.

52. William D. Leahy, *I Was There,* pp. 347-349.

53. Robert H. Ferrell, *Harry S. Truman: A Life,* Norwalk, CT: Easton Press, 1994, p. 191.

54. David McCullough, *Truman,* Norwalk, CT: Easton Press, 1982, pp. 738, 770-771. The National Security Act of 1947, July 26, 1947 reprinted in Alice C. Cole, *et al., The Department of Defense: Documents on Defense Establishment and Organization, 1944-1978,* pp. 35-50.

55. Harry S. Truman, "The Memoirs of Harry S. Truman," Vol. 2, *Years of Trial and Hope,* Garden City, New York: Doubleday and Company, 1956, pp. 442-444.

56. Harry S. Truman, "The Memoirs of Harry S. Truman," Vol. 1, *Year of Decisions,* Garden City, New York: Doubleday and Company, 1955, pp. 7, 17.

57. Dwight D. Eisenhower, "Annual Message to the Congress on the State of the Union," January 1958, *www.c-span.org/executive/transcript.asp?cat=current_event&code=bush-admin&year=1958,* link valid as of July 23, 2003.

58. Dwight D. Eisenhower, "The White House Years," Vol. 2, *Waging Peace, 1956-1961,* pp. 221, 244-247, 251-252.

59. Ronald Reagan, "Message to Congress Outlining Proposals for Improving the Organization of the Defense Establishment," the White House, April 24, 1986, printed in United States, 99th Congress, 2nd Session, House Document 99-209,

Defense Reorganization: Message From the President of the United States, Washington, DC: U.S. Government Printing Office, 1986.

60. *Ibid.*

61. John Sweetman, "A Process of Evolution: Command and Control in Peacetime," in *Sword and Mace: Twentieth Century Civil-Military Relations in Britain*, John Sweetman, ed., London: Brassey's Defence Publishers, 1986, pp. 41-53.

62. Italics in original. Alexander Hamilton, "Number 26," Alexander Hamilton, James Madison, and John Jay, *The Federalist*, Benjamin Fletcher Wright, ed., Cambridge, MA: Harvard University Press, 1961, p. 216.

63. Italics in original. Alexander Hamilton, "Number 69," Alexander Hamilton, James Madison, and John Jay, *The Federalist*, edited by Benjamin Fletcher Wright, Cambridge, MA: Harvard University Press, 1961, p. 446.

64. Michael Les Benedict, *The Blessings of Liberty*, Lexington, MA: D. C. Heath, 1996, pp. 85-86.

65. United States, 108th Congress, Senate, Committee on Armed Services, "Advance Questions for General Richard B. Myers, USAF Nominee for the Position of Chairman of the Joint Chiefs of Staff," July 24, 2003, p. 20, *http://www.senate.gov/~armed_services/statement/2003/July/Myers.pdf*, link valid as of October 21, 2003.

66. United States Congress, Senate, Committee on Armed Services, *Department of Defense Reorganization Act of 1986 Report, To Accompany S. 2295) . . . Together With Additional Views*, Washington, DC: U.S. Government Printing Office, 1986, pp. 2-4.

67. *Ibid.*, pp. 25-26.

68. Chairman of the Joint Chiefs of Staff Instruction, CJCSI 3100.01A, September 1, 1999, "Joint Strategic Planning System" and *Joint Publication 1-02: Department of Defense Dictionary of Military and Associated Terms, April 12, 2001, as amended through June 5, 2003*, Washington, DC: U.S. Government Printing Office, 2003, p. 289.

69. Message to Congress, May 16, 1940, reprinted in Franklin D. Roosevelt, *Roosevelt's Foreign Policy, 1933-1941: Franklin D. Roosevelt's Unedited Speeches and Messages*, New York: Wilfred Funk, 1942, pp. 237-241.

70. Ronald Reagan Remarks at the Annual Convention of the National Association of Evangelicals, Orlando, Florida, March 8, 1983, *The Public Papers of the Presidents of the United States: Ronald Reagan, 1983*, Vol. 1, "January 1 to July 1, 1983," Washington, DC: U.S. Government Printing Office, 1984, p. 364.

71. Ronald Reagan, "Message to Congress Outlining Proposals for Improving the Organization of the Defense Establishment," the White House, April 24, 1986, printed in United States, 99th Congress, 2nd Session, House Document 99-209, *Defense Reorganization: Message From the President of the United States*, Washington, DC: U.S. Government Printing Office, 1986.

72. In 1983, Representative Ike Skelton of Missouri proposed the abolishment of the CJCS and the creation of "a single Chief of Staff for the National Command Authorities." Under Skelton's proposal, the new Chief of Staff would be designated as the principal military advisor to the President, the Secretary of Defense, and the National Security Council. Skelton's proposal would establish a military chain of command that ran through the Chief of Staff. It also preserved officer rotation to the Joint Staff that supported the Chief of Staff and mandated that officers return to their services at a set interval. Testimony of Congressman Ike Skelton and proposed bill, H.R. 2560, "Military Command Reorganization Act of 1983," United States, 99th Congress, House, Committee on Armed Services, H.A.S.C. No. 99-81, *Reorganization Proposals for the Joint Chiefs of Staff*, Washington, DC: U.S. Government Printing Office, 1983, pp. 31-45, 47, 56.

73. "Organization of the General Staff," in Elihu Root, *The Military and Colonial Policy of the United States: Addresses and Reports*, Cambridge, MA: Harvard University Press, 1916, p. 429.

74. In 1947, the members of the Navy and Marine Corps expressed fears that the creation of an overarching General Staff would result in the elimination of the Service Secretaries and their staffs. Largely at the insistence of the Navy, the National Security Act of 1947 prohibited the establishment of an Armed Forces General Staff. The National Security Act of 1947, July 26, 1947, in Alice C. Cole, *et al., The Department of Defense: Documents on Defense Establishment and Organization, 1944-1978*, Washington, DC: Office of the Secretary of Defense Historical Office, 1978, p. 41.

75. Unlike the British Ministry of Defence, the Service Departments would remain autonomous rather than subordinate to the Joint Staff. For details of the organization and composition of the Ministry of Defence, see *http://www.mod.uk/*, link valid as of October 21, 2003.

☆U.S. GOVERNMENT PRINTING OFFICE:2004-604-061/00004